THE

SUSTAINABLE

WAY

Straight talk about global warming - what causes it, who denies it, and the common sense transition to renewable energy.

JAMES LEACH

STA Publications
Email: stapubs@comcast.net

March 31, 2016

Contents

PREFACE

"Sustainable Development is development that meets the needs of the present without compromising the ability of future generations to meet their own needs." - International Institute for Sustainable Development (IISD)

This book grew out of my quest to understand global warming. Like many Americans I have heard a lot of conflicting information about human-caused climate change. Most of it supports the view that it is real and a serious danger to mankind. Some have said it is a hoax, and what trickles in through the news media often creates more questions than answers. We've seen political fighting, questioning of the science, and special interests influencing the debate one way or the other.

I'm not a scientist, just an ordinary concerned citizen who cares about the planet we are leaving to future generations. As I reviewed the climate science literature, I encountered many lengthy and complicated reports written by scientists. I also read numerous articles and books intended more for regular folks like me. All of this took a lot of time and was sometimes tedious and confusing. With our busy lives, it is no wonder that so many people are confused, uninformed, or misinformed about global warming. When I finally came to some conclusions, I realized that it would be useful to have a source of information that offered a concise and plainspoken description of the causes and consequences of global warming, how we arrived at this point of political gridlock, and what options we have for solving it. This is my attempt to fulfill these goals.

My research has led me to conclude that global warming caused by human activity is a serious threat to life as we know it. If you are already of another view, I encourage you to take a breath, commit to opening your mind, and read on to discover what is presented here. Perhaps you will discover some new information or a new perspective. You may even change your mind.

A few words about the science, economics, and politics are in order. Climate science is complex, and includes experts from many countries working in many disciplines. Across the globe, developed nations, like the United States, are responsible for most of the emissions from the use of fossil fuels and other sources that have contributed to global warming. As climate science evolved through the past decades, it became clear to the nations and enterprises involved in providing fossil fuels, that any actions to curtail consumption were potential threats to their wealth, power, and prosperity. It should come as no surprise that addressing global warming would be contentious because it is after all the established order at stake. The resulting political intrigue has obscured the science, created confusion, and helped to avoid meaningful international efforts to literally save the planet for human habitation.

Whose climate science can you trust?

The United Nations Intergovernmental Panel on Climate Change (IPCC) was established by the United Nations Environment Program and the World Meteorological Organization in 1988 "to provide the world with a clear scientific view on the current state of knowledge in climate change and its potential environmental and socio-economic impacts." Scientists from all over the world contribute to the IPCC by providing and reviewing research reports. IPCC periodically publishes assessment reports to communicate the latest scientific knowledge about climate change. Their latest report, Assessment Report 5, was completed in 2014. Copies of all of their reports are available at the IPCC web site (http://www.ipcc.ch). Many United States-based scientists are active research contributors to IPCC. United States government agencies, such as the Environmental Protection Agency, the National Aeronautical and Space Administration, and the National Oceanic and Atmospheric Administration also publish climate change information.

Is the science settled? Is global warming caused by human activity? Is it a serious threat to the environment and mankind? I say YES to all of these questions. Understand that this doesn't mean that every detail is settled, that all projections are going to be exact, or that some findings won't change over time as new knowledge is acquired. Some fields of science enjoy a high degree of certainty because controlled laboratory experiments allow for easy proof of cause and effect. Climate science is a much more complicated situation. Enough of the science is settled to clearly show that human activity is causing global warming and to demonstrate that if we do not do something to stop and hopefully reverse it soon our planet will face catastrophic consequences, if not in our lifetimes, most certainly in the lifetimes of future generations. It comes down to what kind of legacy we want to leave.

Although there is comforting good news about the abundant renewable energy we have at our disposal, much of what follows in this booklet is not what one would call a happy story. It is a rather blunt description of the perilous circumstances mankind has created and how it is that we have failed to do enough to avoid the serious consequences of climate change from human-caused global warming. And because all of us are part of the problem, you will be confronted with the necessity to accept some personal responsibility and for taking actions to help avoid the impending crisis.

OUR LEGACY

"Only within the moment of time represented by the present century has one species—man—acquired significant power to alter the nature of the world." - Rachel Carson, *Silent Spring,* 1962

A Story Told from Late in This Century

There was once a time in the opening decades of the 21st century when most people living in the developed nations of the world enjoyed a life of peace, plenty, and security. Food, clothing, and shelter were in affordable abundance. Countless laborsaving devices freed most individuals from the many burdensome tasks of bygone times. Electronic communications technology not only connected people around the world but also opened access to the collective knowledge of mankind to inform, entertain, and conduct commerce. Transportation technology had evolved to the point where even those of modest means could conveniently travel at speed and in safety. Vast transportation networks on land, sea, and in the air supported highly efficient global commerce, dramatically raising living standards across the globe and fostering cooperation and peace among trading nations.

Driven by abundant energy, this hyper-capitalist world economy produced a period of tremendous trade expansion and economic growth. Assured by the promise of continued advances in technology and human ingenuity, a general feeling of optimism in a future with ever-higher living standards, peace, and security prevailed. This was all made possible because mankind had developed the scientific knowledge to exploit the natural resources this privileged way of life required. Then as the accumulated buildup of unrestrained greenhouse gas emissions began to alter the climate, the citizens of the world became increasingly alerted to the impending harm these continued emissions would cause.

Past the Tipping Point

Faced with the scientific evidence of human-caused global warming, the nations of the world met repeatedly to negotiate policies to curtail greenhouse gas emissions. Some nations took decisive steps to transition away from fossil fuels, to limit emissions, and to reverse global warming. In spite of all the evidence, those in control of the United States Congress remained committed to denying that human-caused global warming existed and to discouraging development and use of renewable energy to replace fossil fuels. Fearing economic disadvantage, many other nations also delayed their efforts to limit greenhouse gas emissions. As the world population grew and more people sought higher livings standards, the energy needs of the world continued to rapidly grow, dramatically accelerating greenhouse gas levels and pushing temperatures even higher.

It was hotter on average in many places, colder in others, and more extreme weather events occurred everywhere. Some areas developed persistent drought conditions, while others had increased precipitation. Long established climate norms changed, altering growing seasons, bird migrations, and animal habitats. Pollution from fossil fuel extraction, production, and consumption became increasingly serious causing evermore damage to the environment, premature loss of life, and human suffering. As the melting of the world's glaciers and permafrost soils accelerated, sea levels rose rapidly and enormous quantities of greenhouse gases were released into the atmosphere. All the while the oceans progressively lost the capacity to buffer global warming and to sustain life.

Gradually, as the perilous circumstances became obvious, all the major nations came together to address the crisis and to mitigate the impending consequences. But it was too late. The tipping point had passed. The warming continued to accelerate, pushing climate conditions in many parts of the world to uninhabitable extremes. Formerly fertile areas across the globe became unproductive. Wildfires ravaged the parched land. Rising oceans flooded low islands and coastal zones, including many of the most populated and productive agricultural lands and commercial hubs. The changes to the world's oceans destroyed coral reefs and other aquatic conditions essential to the marine food chain. Countless life forms living on the margins of existence and unable to adapt or move to livable habitat went extinct. Massive storms produced unprecedented widespread flooding and damage. Shifting oceanic and atmospheric currents dramatically altered regional climatic norms that had existed for all recorded history.

As the Atlantic Gulf Stream faltered, Northern Europe cooled dramatically. The rising ocean flooded most of the Eastern and Southern Indian subcontinent making it uninhabitable. In North America, the rising sea took nearly all of developed Florida and the low-lying coastal zones. Major cities such as New York, Boston, Charleston, New Orleans, and Washington D.C. became inundated by the intruding waters. Without sufficient mountain snowpack, the agricultural areas of Western North America withered. The grain growing regions of the world failed to produce reliably. Extreme high summer temperatures dramatically reduced human productivity and caused unprecedented suffering and loss of life across the globe. Food, water, and many of the other basics of life became far more expensive, and for many unaffordable or unavailable. Scarcity and dislocation bred conflict and persistent violence across the globe. Hundreds of millions of people died of starvation and in violent struggles. Countless more became refugees. The integrated global economic bubble burst, destroying fortunes worldwide and undermining security and peace for everyone. The forward trajectory of growing opportunity and prosperity was reversed. The Earth was no longer a bountiful place to live a joy-filled life.

This Is No Fantasy

The increasing accumulation of atmospheric greenhouse gases, primarily from the use and production of fossil fuels, has put in motion the disaster described in this scenario. Although the tipping point leading to an irreversible catastrophe may not have yet occurred, the overwhelming weight of scientific evidence clearly shows that mankind's fate is grim if we fail to curb global warming. These consequences are not imagined; all are already occurring someplace on the planet, even if not yet on the scale that will ultimately occur if we fail to act soon. This outcome is not inevitable; it can be averted.

Catastrophe Averted – An Alternative to Passing the Tipping Point

During the second decade of the 21st century, as knowledge of the potential consequence from global warming became more broadly understood, the citizens of the United States elected government officials committed to reversing global warming and a transition to renewable energy. All of the nations of the world came together to put in place aggressive plans to slow and then reverse global warming. Greenhouse gas emission regulations were applied to provide certainty to industry and to unleash the power of competitive innovation to most effectively accomplish the transition. These efforts included carbon cap-and-trade programs, greenhouse emission fees, and credits to establish clear goals and incentives to accelerate the shift to renewable energy. Substantial government and private investments in research and infrastructure accelerated the renewable energy revolution.

Worldwide as the consequences from the existing accumulation of greenhouse gases played out, climate conditions continued to change, altering average temperatures and precipitation norms. Everywhere people were forced to adjust to the altered conditions. Many could cope, especially those with more resources. Some locations justified extensive mitigation investments to protect valuable assets, such as sea walls and levies. Many coastal areas and regions could not be saved and were abandoned. In some instances violent conflicts over competition for scarce resources could not be avoided.

In the United States as dependence on fossil fuels eased, the long-standing transfer of wealth to energy rich foreign nations and cartels ceased. Increasingly detached from the dependence on imported fuels, the United States was no longer forced to deploy military forces across the globe to protect foreign suppliers and to prop up their governments. Unavoidable costs were associated with population relocations, damage from severe weather, and mitigation efforts. The investment in renewable energy, the influx of affordable renewable energy, and the reduced environmental and societal costs from reduced fossil fuel use produced a strong economy with broadly shared prosperity.

As the world population grew and more people sought higher livings standards, their energy needs were increasingly supplied by renewable sources. Greenhouse gas emissions eased, causing a gradual shift downward in the global warming trend line. Pollution from the extraction, production, and consumption of fossil fuels dramatically declined, reversing the levels of environmental damage, premature loss of life, and suffering all had previously been exposed to. By mid-century the major industrial nations were well on their way to a transition to carbon-free or carbon-neutral status and increasingly taking advantage of abundant and affordable renewable energy.

Because of the aggressive steps taken early in the century the long-feared tipping point of catastrophic climate change was averted. The atmosphere and oceans began the gradual restoration to the conditions all life had long depended on. The integrated global economy continued to prosper, ensuring security, opportunity, and peace for people everywhere. The Earth was an even more bountiful place to live a joy-filled life

And This Is No Fantasy

The Earth is awash in untapped renewable energy that can fully meet all of mankind's energy needs many times over. Though the transition will require significant investments and time to scale up to low-cost production levels and to build out the required infrastructure, we already have the know-how and technology to fully replace fossil fuels. Waiting any longer to get started simply isn't beneficial. Of course, there will be useful breakthroughs and also some disappointments along the way as some technologies or methods for capturing the energy prove better than others. And there will also be gradual adjustments we will all need to make as the transition takes place over decades. The most important realization is that this renewable energy revolution will not only provide abundant affordable energy, it will also produce continued economic vitality, a more broadly shared prosperity, and a far healthier environment for all life forms.

The most urgent priority here in the United States is to break the controlling influence the fossil fuel industry has exerted over so many of our elected officials. This will require a far more informed and motivated electorate that is dedicated to addressing global warming. Though some negative consequences from the existing buildup of greenhouse gases are unavoidable and will require some adaptions and mitigation, if we unite soon with the other major industrialized and developing nations of the world to earnestly act to solve this most serious threat, we can avoid the catastrophe and leave a legacy to be proud of.

Now let's consider the greenhouse gases and their impact on climate.

THE GLOBAL THERMOSTAT

"In nature nothing exists alone." - Rachel Carson, *Silent Spring*, 1962

The Earth receives enormous amounts of energy from the sun. Much of it escapes our grasp as it is reflected or radiated back into outer space. Most of the energy that is not reflected is either used in natural processes, like those that produce plant life through photosynthesis, or is absorbed to heat the land, oceans, and atmosphere. This absorbed heat as well as heat from other sources, such as from the burning of fossil fuels, is constantly radiated through the atmosphere toward outer space.

Our atmosphere contains a relatively uniform combination of gases that varies slightly depending on the amount of water vapor in it. It consists of from 75% to 78% nitrogen, 20% to 21% oxygen, water vapor ranging from as much as 4% to 0%, close to .9% argon, and about .1% containing small quantities of many other gases. A small portion of these gases absorb heat. This heat absorbing quality is a primary mechanism that regulates global temperature. Reduce the heat absorbing gases and the temperature cools; add more and it warms. (NOAA Greenhouse Effect) The gases responsible for nearly all of this heat absorption are water vapor, carbon dioxide, methane, and nitrous oxide.

Although water vapor packs a potent heat absorbing effect, it is not considered a gas that causes global warming. The amount of water vapor in the atmosphere is instead a *climate feedback* directly related to temperature; the warmer the atmosphere, the more liquid water evaporates and the more water vapor the atmosphere holds. The level of water vapor is therefore considered an *effect* of climate temperature and not a *cause*. (NOAA Water Vapor) The gases that are responsible for practically all global temperature regulation, what are called "greenhouse gases," include carbon dioxide, methane, nitrous oxide, and various man-made fluorinated gases. (NOAA Greenhouse Effect) The amount of these gases in our atmosphere depends on both natural processes and human activities that release and remove them. Collectively, the greenhouse gases presently account for only about .04% of the atmosphere. (IPCC AR5, 2014) When you consider just how small a portion this is (i.e., .04% = 4/10000 = 1/2500), it becomes easier to grasp how it is that large-scale emissions of these gases from human activity can change the climate.

Climate science research has shown that greenhouse gas levels had long remained stable until the onset of the Industrial Age. As you can see in Figure 1, atmospheric carbon dioxide (CO_2), methane (CH_4), and nitrous oxide (N_2O) increased with industrial-scale burning of fossil fuels including coal, oil, and methane. Between 1880 and 2012 the globally averaged combined land and ocean surface temperature warmed 1.53 °F (0.85 °C). Although the industrialized nations have been the primary producers of these emissions, the consequences of this "greenhouse effect," are felt in various ways everywhere on the planet. (IPCC AR5, 2014)

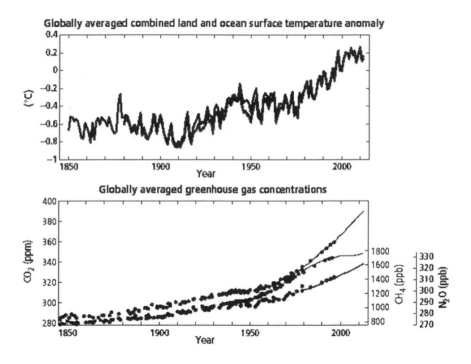

Figure 1: Globally Averaged Combined Land and Ocean Surface Temperature Anomaly (IPCC AR4, 2014)

Predicting the level of future global greenhouse emissions and the effect on climate is an uncertain task. It depends on many variables, including how mankind will respond to the challenge. Modern climate simulation technology has provided scientists with powerful tools to predict average global temperature. To more clearly show the effect from different greenhouse gas emission scenarios, researchers have developed several models or *adaption pathways* ranging from a Low Emission Model that assumes we do a great deal to curb emissions, to a High Emission Model that assumes we do little. The predicted average U.S. temperature increase through the end of this century on the Low Emissions Model is 3.0 °F (1.7 °C); the High Emission Model is 12 °F (6.7 °C). (EPA U.S. Climate Models)

The U.S. can likely cope with a 3.0 °F increase, though not without experiencing significant economic damage, environmental degradation, dislocation, loss of life, and other suffering. The impact will vary depending on location and on the resources available. Many, especially the poor, will likely not cope as well as others. And what about the course we are presently on, the High Emission Model with a 12 °F increase in average temperature? No one is saying we can easily cope with such an increase! We need to start acting now to prevent this catastrophic outcome.

Carbon Dioxide and the Carbon Cycle

Carbon dioxide is the most prevalent greenhouse gas and is currently responsible for about 76% of the greenhouse effect. (IPCC AR5, 2014) (Mann and Kump, 2015) It is a clear, odorless, nonflammable gas that we tend to think of as harmless, something that merely makes our carbonated beverages fizzy.

The change in atmospheric carbon dioxide over time depends on the balance between the emissions of carbon dioxide from all sources and the amount of carbon dioxide that is removed, what is commonly called the carbon cycle. The amount of carbon dioxide in the atmosphere has increased by 38% since 1850. As shown in Figure 1, the rate of increase has been gradual but steady through the mid-1950s when it took a sudden upturn tracking the growth in fossil fuel consumption. (IPCC AR5, 2014)

Where Does Carbon Dioxide Come From?

A primary source of atmospheric carbon dioxide is the burning of fuels containing carbon, such as wood, peat, coal, and methane, and petroleum fuels such as oil, propane, gasoline, and diesel. These various fuels contain different levels of carbon and other elements, such as hydrogen. As they interact with oxygen during combustion some produce less carbon dioxide emissions for a unit of heat than do others. This is why coal, which has high carbon content, is described as having high carbon dioxide emissions, while methane is described as a relatively low carbon dioxide emitter producing about one-half as much carbon dioxide for a unit of heat. (U.S. Energy Information Administration)

The Earth naturally expels significant amounts of carbon dioxide through volcanic fissures and hot vents in its crust. Carbon dioxide is also formed from the decomposition of organic matter where oxygen is present, such as in forests, open fields, exposed peat soils, and plowed agricultural land. And because underground fossil fuel deposits often contain large amounts of carbon dioxide, natural seeps from these deposits and extraction operations often release significant amounts of carbon dioxide. Ice and permafrost soils that have trapped carbon dioxide within their water crystals also release carbon dioxide as they melt. Surprising to many, cement production is responsible for about 5% of total carbon dioxide emissions through the processing of calcium carbonate (i.e., limestone) to produce calcium oxide (i.e., lime). (Emissions from Cement Industry)

Where Does Carbon Dioxide Go?

Significant levels of carbon dioxide are removed from the atmosphere by plants and other living organisms through photosynthesis. This process breaks down the carbon dioxide molecules drawn from their surroundings, releasing oxygen and extracting carbon for use in building the organisms' cells. This is the beginning of the food chain, passing carbon on to each successive consuming organism. All living cells and the remains of all formerly living things, what we call organic matter, contain carbon. (NOAA Carbon Cycle Science)

More than 50% of the photosynthesis on the planet takes place in our oceans. (EarthSky) As the remains of the photosynthetic ocean plants and other life forms settle to the seafloor, deposits containing carbon are formed. As the seafloor interacts with the shifting plates that make up the Earth's outer crust and with changes in sea level, some of the deposits formed on the seafloor end up on dry land. Limestone is formed in this way. Ocean water also directly adsorbs carbon dioxide making the water more acidic. (Scripps Institution CO2) The same is true of water vapor as it absorbs carbon dioxide forming the carbonic acid that is a normal component of rain and snow. Where seafloor plate subduction occurs, carbon deposits on the sea floor and ocean water containing carbon dioxide are drawn down under the Earth's surface. Permafrost soils and ice also trap the carbon dioxide contained in their source waters. Carbon dioxide has been used for years to inject into petroleum fields to increase production. (US Dep Energy CO2 EOR) Some of it undoubtedly remains trapped there.

All fossil fuels were formed from the remains of organic matter. (U.S. Department of Energy Fossil Energy Source) Collectively, all of the Earth's deposits of these fuels consisting of coal, oil, and methane represent a vast storehouse of carbon. Recent research reveals that the carbon content of the known reserves of these fuels are so extensive that they far exceed the planet's capacity to burn them without pushing global temperatures to catastrophic levels. As much as 66% of these reserves will never be extracted and consumed as fuel because the warming affect would simply make the planet uninhabitable for human life as we know it. This revelation is gradually being recognized in financial markets and energy investment strategies. (UWOSH Carbon Bubble) Sophisticated investors and those managing many of the largest endowment funds are already divesting their fossil fuel holdings. The challenge for this industry is no longer who can find and produce the most fossil fuel. Instead it is who will supply and profit from the limited amount that will be allowed to be produced. (350.ORG) (Carbon Tracker Unburnable Carbon)

Bringing down the level of atmospheric carbon dioxide will require that more carbon is removed through the natural carbon cycle and by other means than is emitted.

Methane's Role

Atmospheric methane is a colorless, odorless, flammable gas presently responsible for about 16% of the greenhouse effect. It is the primary ingredient in natural gas. Although there is far less methane in the atmosphere than carbon dioxide, by volume it is much more effective than carbon dioxide at absorbing heat. (US Environmental Protection Agency) The atmospheric concentration of methane has increased 125% since 1850. (IPCC AR5, 2014) (Mann and Kump, 2015)

Where Does Methane Come From?

Methane is produced from the decomposition of organic matter when little oxygen is present. Natural sources of methane include animal and microbial food digestion, organic decomposition from forests, wetlands, and oceans, seeps from land and ocean fossil fuel deposits, and releases from melting ice and permafrost soils. Human activity related emission sources include organic decomposition from landfills, waste treatment activities, grazing animals and manure management, rice agriculture, bio-fuel production, fossil fuel exploration and extraction, and especially from leaks through the distribution, storage, and use of natural gas. Human activity is currently responsible for 60% of global methane emissions. (U.S. Environmental Protection Agency)

Where Does Methane Go?

Enormous deposits of methane exist in the oceans at great depths trapped in frozen water crystals and in the Earth's crust in areas associated with fossil fuel deposits. Considerable volumes of methane are also trapped in frozen permafrost soils, peat bogs, and wetland soils. (U.S. Department of Energy - Hydrate) Although natural interactions in the atmosphere and soil gradually break down methane, this control is easily overwhelmed when excessive levels are released. (U.S. Environmental Protection Agency)

Nitrous Oxide and the Nitrogen Cycle

Atmospheric nitrous oxide is a colorless and slightly sweet smelling, nonflammable gas now responsible for about 6% of the greenhouse effect. Though the amount of nitrous oxide in the atmosphere is small compared to methane, by volume it is far more effective at absorbing heat. (US Environmental Protection Agency) The atmospheric concentration of nitrous oxide has increased 20% since 1850. (IPCC AR5, 2014) (Mann and Kump, 2015)

Where Does Nitrous Oxide Come From?

The major source of atmospheric nitrous oxide is from the natural processes that break down nitrates, the primary ingredient in fertilizers such as ammonium nitrate. Nitrates are formed naturally from nitrogen by microbial and chemical interactions in soil, the seas, and in animal digestive systems. Nitrates supply nitrogen to land and marine plants. When plants do not take up nitrates, these natural processes produce nitrous oxide. Because much of the Earth's soil is lacking sufficient natural sources of nitrates, modern large-scale agriculture has come to depend on commercial nitrate fertilizers. Over-application or poorly controlled application of nitrate fertilizers is the primary man-made source of nitrous oxide. Other human-caused sources of nitrous oxide include burning of fossil fuels, soil cultivation, waste from livestock production, waste treatment facilities, and industrial processes involving production of synthetic fiber products. (U.S. Environmental Protection Agency)

Looking only at the United States between 1990 and 2013, 75% of nitrous oxide emissions from human activity came from agricultural soil management, 5% from manure management, 11% from fossil fuel burning, and 9% from industry and chemical production and other sources. On our present course, nitrous oxide emissions are expected to continue increasing in coming years mainly from agricultural sources and burning of fossil fuels. (U.S. Environmental Protection Agency)

Where Does Nitrous Oxide Go?

Small amounts of nitrous oxide are stored in soils. In the atmosphere, it is gradually destroyed through a reaction with light and oxygen and has an average life of 114 years. (Center for Climate and Energy Solutions)

Man-Made Fluorinated Gases

The fluorinated gases we are most commonly aware of are those that are used as refrigerants in heating and cooling systems and as propellants in spray cans. Though the amount of fluorinated gases in the atmosphere is small, they are exceptionally long lasting and extremely effective heat absorbers. Fluorinated gases currently account for about 2% of the greenhouse effect. (IPCC AR5, 2014) (Mann and Kump, 2015)

Many of these gases are destructive to the ozone layer that serves to block harmful ultraviolet light from penetrating the Earth's upper atmosphere. An international agreement from the late 1980s known as the Montreal Protocol and other international agreements act to regulate and limit the use of these ozone-layer depleting gases with the eventual aim to eliminate their use. In spite of these agreements, levels of fluorinated gases have increased in recent years.

Climate Interactions, Tipping Points, and Feedback Loops

Climate change isn't just about the greenhouse gases. Many environmental interactions also have a big impact. Some of these alter the climate in some way. Most are gradual, though some are best described as having a tipping point, meaning something builds to a point then causes a sudden change. Feedback loops also exist, such as a positive feedback loop in which a change produced through a process contributes to future change to the same process, making future changes even more consequential, amplifying or accelerating the rate of change. (NASA Earth Observatory)

Global Warming Is Not Evenly Distributed or Consistently Experienced

It is often said that the biggest factor preventing the general public and perhaps some of our political leaders from believing in global warming is the variability of local climate. (NASA Public Perception of Climate Change) It is easy to see how local weather events that appear out of line with a warming climate, such as an unseasonal cold spell or a prolonged winter, can cast doubt on the concept. Weather is complicated. Prevailing winds, ocean currents, the shape and location of the continents, microscopic airborne particles, and the complex interactions driven by differences in temperature, density, and water content, all stirred up by the tilted rotation of the planet, combine to create the sometimes inconsistent and unseasonal conditions we all experience from time to time. Doubting what climate science tells is based only on one's day-to-day local impressions or the occasional weather anomaly is a serious mistake when so much hard evidence exists that global warming is real.

Climate scientists speak in terms of average global temperature change over some period of time. Such forecasts do not mean that all locations on the planet will feel the same average change. The changes will be different closer to the poles than at the equator. The air over land will typically be warmer than over water. In urban areas the concentration of heat retaining building materials, more waste heat sources (i.e., from fuel combustion, friction, and so on), and less cooling from water evaporation, all combine to make these "heat island" areas hotter than surrounding rural locations. Equatorial areas with warmer climates and more atmospheric water will likely feel the warming effects more and suffer more from increased storm severity.

So when you consider the temperature change you may experience resulting from a change in the average global temperature, remember that it will be different depending on where you are. It may be less, or a lot more, and you won't be capable of casually perceiving the slight difference from one year to the next. While every weather event can be said to be influenced by global warming, no single event can be attributed only to it or be used to refute it; only the long-term trends reveal the truth. Fortunately, many dedicated researchers have been documenting the hard facts. The numbers don't lie. You can trust the long-time scale trends found in their statistics. (IPCC AR5, 2014) (Romm, 2015)

Another factor that is preventing many people from taking global warming seriously is that the immediate effects are often happening somewhere far away and many of the early signs are hard to personally experience. Few of us live in a polar region where some of the most obvious effects are occurring. How many of us depend on a nearby coral reef that is dying or live on an island or seashore that is slowly being swallowed by the sea? And what if the ocean becomes warmer and more acidic? Does it really matter to me if it felt fine on my last vacation? There is no shortage of books and reports on the subject and stories in the news, and yet so many still aren't convinced. The problem is that if we wait until the warming is an obvious "in your face" runaway process, it will be too late.

Habitat Destruction

Anything that reduces the amount of photosynthesis also reduces the amount of carbon dioxide that is removed from the atmosphere. Although plants depend on carbon dioxide, they must also have the other essential ingredients of life, such as an appropriate environment in which to grow, water, minerals, energy, and so on. Habitat loss reduces the amount of carbon dioxide processed through photosynthesis. Deforestation and unsustainable forest practices as well as overgrazing and poor agricultural land use practices are major factors responsible for reducing the amount of photosynthesis. Other human activity such as clearing land for residential, commercial, and transportation space, as well as taking land for water storage, and mineral and fossil fuel extraction also contribute to habitat loss. And as global temperatures rise, more land area will become arid, further reducing the amount of carbon dioxide absorbed through photosynthesis and making the planet still hotter, a good example of a positive feedback loop. (Romm, 2015)

As land-living creatures, we tend to recognize our loss of plant habitat on land. What about the other 70% of the planet's surface covered by the oceans? Increased greenhouse gases are producing significant changes in our oceans. Most of the additional accumulated heat from global warming, about 90% of it, is now stored in our oceans. Not only do increased greenhouse gases directly alter our climate, it also affects organisms living in the oceans. Microscopic marine plants, what are collectively called phytoplankton, are responsible for more than half of the planet's photosynthesis. As atmospheric carbon dioxide levels rise, the oceans will absorb more carbon dioxide and become more acidic, eventually reaching a point where little more can be absorbed. (U.S. Environmental Protection Agency - Oceans) (Mann and Kump, 2015) The increased temperatures and increased acidity interfere with marine life such as corals, shellfish, and phytoplankton. Increased acidification not only blocks their uptake of calcium carbonate to form their structures, it also interferes with their ability to access essential trace metals, such as iron and zinc, and minerals such as nitrogen. (Princeton University, Climate and Energy Research Challenge)

Ocean warming and acidification represent the potential for habitat destruction on a grand scale. This threatens half of the planet's ability to remove atmospheric carbon dioxide and puts the entire ocean food chain in jeopardy. The oceans are already 25% more acidic than they were in preindustrial times. (IPCC AR5, 2014) At what point will all the corals and shellfish die? Will the entire marine food chain collapse? Will the carbon dioxide saturation point be reached before the acidification tipping point triggers a massive phytoplankton die off? Will a large-scale phytoplankton die off create a positive feedback loop that suddenly accelerates increased atmospheric carbon dioxide sending temperatures soaring? Although we really don't yet know enough to answer all these questions, we do know that if we don't begin soon to take the necessary steps to reverse the increased acidification and warming of our oceans, we will someday be facing a true calamity.

Water

Because water vapor is such a strong heat absorber, it packs a powerful climate warming effect. (NOAA Water Vapor) A warmer planet will evaporate more water and hold more water vapor in the atmosphere, thereby absorbing yet more heat and triggering a positive feedback loop trapping even more heat, and evaporating still more water. More water vapor may also cause more cloud cover. Because clouds reflect some of the sun's energy back into space preventing it from reaching the Earth's surface, clouds represent a cooling negative feedback. Although the overall heating effect from increased water vapor may be somewhat offset by the additional cloud cover, having an atmosphere containing much more water is a serious hazard, not only because of its powerful warming effects, but also because of the increased drought conditions, desertification, and the extreme weather it will produce. (NASA Earth Observatory)

Ice- and snow-covered surfaces reflect much more of the sun's energy back toward outer space than do darker colored surfaces, such as plants, exposed soils, or liquid water. With increasing global temperatures, ice- and snow-cover on land and the oceans will be reduced making the resulting less reflective surfaces absorb more heat. This is the same effect we all have experienced with dark-colored clothes being hotter in the sun than light-colored clothes. It will also trigger the release of truly enormous amounts of greenhouse gases such as carbon dioxide and methane from melted permafrost soils and ice, another powerful positive feedback loop.

A major factor contributing to sea-level rise is ocean warming. As water warms, it expands to occupy more space. (IPCC AR5, 2014) (Romm, 2015) The melting of land-based ice and snow will also significantly raise sea level, just as adding water or ice cubes to an already full glass of water will cause the glass to overflow. Of special concern are the massive glaciers of the Antarctic. The ice shelves on the sea next to these glaciers have been showing signs of breaking off from the continent. Climate scientists have been reporting and warning about this for many years. Just as they have predicted, several major ice shelves have already broken off in recent years. Because some of these shelves act to hold back the glaciers on the continent, it is expected that their disintegration will set in motion flows of these most massive glaciers into the sea. Other large glaciers, such as those on Greenland, have already been melting and increasingly flowing toward and into the sea. (Romm, 2015)

Unlike the melting of ice and snow on land, the melting of ice formed on ocean surfaces will not raise sea level, just as a melting ice cube floating in a full glass of water does not cause the glass to overflow. However, significantly reducing the amount of ice forming on ocean waters will disrupt the heavy saline water undercurrents that are generated as the polar ocean waters freeze forcing out salts into the waters below. These undercurrents of heavy saline water help drive the ocean's major current systems that serve as conduits for moving heat and cold around the planet and are responsible for creating many regional climate norms that have existed for eons. The Atlantic Gulf Stream is one of the best-known examples of such currents.

How much will the oceans rise? Like temperature changes from global warming, sea-level change will not be the same everywhere. Tidal forces, natural subsidence (i.e., sinking land), local water temperature, and other forces will produce different experiences around the globe. The average global sea level has already increased 8 inches (20 cm) since 1880. The Low Emission Model pegged sea-level rise at 15.75 to 24.8 inches (40 to 63 cm) by the end of this century; the High Emission Model predicts as much as 32 inches (82 cm). (Ellie Zolfagharifard for Daily Mail) (IPCC AR5, 2014) Because climate scientists have since pointed out that these prediction models, which are now a few years old, did not include some recent information showing some increased glacier melting, these estimates may significantly understate seal-level rise for both models. (Romm, 2015)

Atmospheric Aerosols

The atmosphere contains tiny particles called aerosols that act to reflect and scatter sunlight, generally cooling our atmosphere. The light-scattering qualities are most noticeable as they cause spectacularly colorful sunrise and sunset displays. Aerosols come from both natural and man-made sources. The primary natural sources are from volcanic eruptions, desert dust, ocean marine life, and natural fires. The main human sources are man-made fires and burning fossil fuels. Not only do aerosols directly reflect and scatter sunlight, they also interact with chemicals and clouds. For example, sulfur compounds produced from ocean phytoplankton are essential in cloud formation. This is another example of a positive feedback loop where damage to ocean marine life from increased acidification and warming may reduce cloud formation and, therefore, accelerate global warming. Dust aerosols, which absorb heat, also hamper cloud development by preventing condensation, yet another positive feedback.

In the upper atmosphere some aerosols react to break down the ozone layer by allowing more harmful ultraviolet light to reach the Earth's surface. Other aerosols interact with clouds to increase their light reflective qualities. Aerosols from large-scale volcanic eruptions can cause prolonged cooling periods sometimes lasting years. Though some of the effects from aerosols are global, most are felt regionally. (NASA Facts about Aerosols) (Mann and Kump, 2015) (Romm, 2015) Aerosols are yet another reason why weather conditions where you are may differ from the global average.

Now that we have covered the basics climate science, let's examine how the United States got to this point without taking meaningful steps to deal with the global warming crisis.

THE FORCES OF DENIAL

"My point is, God's still up there. The arrogance of people to think that we, human beings, would be able to change what He is doing in the climate is to me outrageous." – James Inhofe, Chairman U.S. Senate Committee on Environment and Public Works

Denial: the refusal to accept reality.

For many decades, climate scientists have been sounding the global warming alarm. In spite of this, not nearly enough action has materialized in the United States to address the problem. Science has provided the foundation knowledge for all the advancements behind our modern way of life. The world embraces, even celebrates science and all it has allowed us to understand and accomplish. So why do we encounter so many people in denial of what climate science has revealed to us? Who are these deniers?

- **Uninformed:** The uninformed deniers simply haven't been exposed to what causes global warming and the impact it will have on climate. These deniers are similar to those who continued to believe tobacco smoke or the poisonous neurotoxins coming out of our tail pipes from leaded gasoline were harmless long after the scientists proved otherwise and their findings were widely publicized.
- **Science Skeptics:** These deniers have come to reject science because it contradicts their perceptions or beliefs. They typically have little practical knowledge of science and find the processes leading to global warming hard to comprehend.
- **Short-Term Opportunists:** These deniers may also be described as "postponers." They may realize that something needs to be done about global warming someday, but believe the consequences to be small, or to be so far off into the future that it doesn't make any sense to act now and ruin a good thing. We are all guilty to some extent of this kind of short-term thinking that rationalizes postponing what we know is right in the long-term to take advantage of short-term opportunities or to avoid inconvenience.
- **Selfish:** The selfish deniers are those who don't have empathy for others and therefore have no concern about issues that don't seem to touch them personally. These deniers simply aren't wired to care about harm caused to

others, especially distant and future others. Unfortunately, many people drawn to "leadership positions" display this personality characteristic. (Forbes.com -Susan Adams - Why Selfish Leaders Triumph, 2011)

- **Hired Guns:** These are deniers who will do or say just about anything if the money is right. They come in many forms including: lawyers, public relations firms, executives, scientists, lobbyists, the media, and of course some politicians.

Externalities

Economists define *externalities* as costs that producers or consumers pass on to others. Externalities result in a transfer of wealth from those who will ultimately pay the cost, to those benefiting from the externality. History is full of examples of externalized costs. In the environmental realm, we have examples of public and private organizations dumping their toxic byproducts into our water and air. Externalities aren't just limited to large organizations. Individuals who leave their trash along the roadside are also imposing a cost on someone else. Externalizing a cost is really thievery by another name. The unfortunate reality is that we humans tend to externalize costs if we can get away with it. Sometimes we think it won't be noticed or attributed to us. Other times we know that it will but we have the power to get away with it. Typically our behavior is linked to how an external cost might come back to harm us. Where there is little or no apparent *direct connection* to us, or if the cost is in the *distant future*, we tend to ignore the issue.

Many decades ago, early in climate research, scientists recognized that carbon dioxide emissions from burning fossil fuels would cause the atmosphere to warm. (Suzanne Goldenberg M. J., 2015) (Suzanne Goldenberg T. G., 2015) The fossil fuel industry realized that capturing these emissions would be difficult and expensive, if not impossible. They knew that some uncertainty existed in the climate scientists' forecasts and that decades would pass before the warming affects would start to have serious noticeable consequences. And of course, they also realized that no immediate alternative energy source was available that could deliver the low cost and convenience that consumers expected. Meanwhile, there was a lot of money to be made. And so the practices by many in the fossil fuel industry to discredit climate science and to block efforts to constrain carbon dioxide emissions began. (Suzanne Goldenberg M. J., 2015) (Katie Jennings, 2015)

The stakes for the fossil fuel industry are enormous. The companies and nations whose wealth is largely represented by their fossil fuel reserves are faced with the reality that if greenhouse gas emissions are to be significantly reduced, fossil fuel use must be drastically reduced, perhaps even stopped completely. Most of their reserves will become stranded assets, and their production and distribution infrastructure and expertise made worthless. The world's most wealthy energy rich nations and companies are fighting to block efforts to curb fossil fuel use because maintaining their immense wealth depends on it. (UCS - Internal Fossil Fuel Industry Memos, 2015)

Regardless of how you feel about the fossil fuel industry, it is impossible to escape the fact that all of us and our predecessors over the past few hundred years have been guilty of externalizing the cost of the emissions that occurred during the exploration, extraction, production, distribution, and burning of these fuels used to deliver the products we use and to provide us with energy we directly consume. If we are looking for someone to blame, it isn't just the fossil fuel industry. We must look in the mirror because it's really all of us.

The Gated Community World Order

History reveals many examples of the powerful imposing their will on others, often taking away freedom and property, causing untold destruction, suffering, even death. One can't help but wonder how those at the top who own and run the fossil fuel and other greenhouse gas producing industries expect to escape the effects of global warming. They know the science, and yet so many deny it. As mentioned earlier, we've seen this before in countless instances of companies found guilty of knowingly selling dangerous products, such as leaded gasoline and tobacco to name only a few. We now know that the leaders and owners in those industries understood how damaging their products were, yet they fought hard to preserve their businesses. Why? It comes down to preserving shareholder value. Fossil fuel industry executives are paid to protect the value of their company and if they don't, they get fired. Under the relatively short-term mindset that prevails in these situations, doing the right thing for mankind just isn't a factor.

Consider the wealthy petroleum investors or petroleum industry executives who believe their personal fortunes are at stake in keeping actions to curb greenhouse emissions at bay. Do they believe they'll be safe living away from where the effects are the worst, hidden behind a figurative gated community? We see this among many of the wealthiest who live in a separate society away from all us regular folks. Their children attend private schools; they can afford the best healthcare, have plenty of food, travel in private jets, and have all the things, including political influence, their wealth can buy. Only when something threatens their security do they take an interest. For the executives and investors of greenhouse gas emitting enterprises, shutting down their emissions threatens their near-term security more than the long-term perceived threat from global warming.

The gated community metaphor also applies to nations. As the effects of global warming become apparent, wealthy nations will likely cope with the initial environmental and physical damage better than poor nations. Do our global warming denying leaders and the fossil fuel elite who support them believe they are insulated from the climate effects felt by most of the ordinary people of their own nation and other less wealthy nations?

The Fox Guarding the Henhouse

Throughout the world, ruling elites are influential in controlling governments and their policies. In countries such as Saudi Arabia, they are easy to identify. The United States has a wealthy elite class of political contributors, busy trying to control who is elected, and paying for influence over government policy. Though politicians from both major political parties have been willing recipients of fossil fuel contributions, the Republican Party has been the most receptive. Today the overwhelming majority of Republicans in the United States Congress are global warming deniers. Senator James Inhofe of Oklahoma, author of *The Greatest Hoax: How the Global Warming Conspiracy Threatens Your Future*, has been one of the most outspoken. The book is a replay of the tactics the industry has used to obstruct both the recognition of global warming and any action to address it. It isn't surprising that Senator Inhofe is such an ardent denier. After all, he once worked drilling oil wells, and represents a state with many people who own wells and many thousands more who work in the industry. Oil and natural gas have been the source of much of Oklahoma's wealth and identity. Mr. Inhofe says that carbon dioxide emissions are harmless and that only God is capable of changing the climate. In his opinion, curtailing fossil fuel use will only mean economic deprivation to his constituency without solving any problem - "All pain for no environmental gain" as he says. He has received substantial contributions from fossil fuel energy interests throughout his political career, including individuals, lobbyists, and Political Action Committees associated with the likes of Koch Industries, Murray Energy, ConnocoPhillips, ExxonMobil, and many more. (Natural Resource Defense Council - James Inhofe)

Together the congressional deniers like Senator Inhofe have attempted to block efforts to reduce fossil fuel use and to prevent a transition to renewable energy. Are these politicians in the pocket of the fossil fuel industry or do they really believe global warming is a hoax? The scientific evidence is there. Are they not willing to read it? Are they unable to understand it? Or are they simply acting at the direction of their benefactors? It is hard to believe that the big money being thrown around in this battle isn't corrupting.

Psychology and politics are also factors to consider. Behavioral research has shown that when confronted with anything that conflicts with our established belief or a belief we adopt from a group we identify with, all of us unconsciously tend to first look for information to confirm our belief rather than to consider the conflicting idea. (Haidt, 2012) Sometimes rational thought can break through this "confirmation bias," but most often it doesn't, especially with people unwilling to open their minds and take the time to investigate. Also, the longer a person holds a belief and the more public it is, the harder it is to change. With Inhofe and the others so committed to climate science denial, it isn't easy for them to change their minds and still look competent. On the political side of the issue, if the voters require climate science denial as a condition for their vote, this will be their elected politician's position.

Who is really responsible for the obstruction and inaction of these politicians? Ultimately a more informed electorate needs to demand that their representatives accept the science and take real actions to solve the problem. Again, look in the mirror and at your friends, neighbors, and fellow citizens.

Partisan Politics

Today's polarized political climate in the United States is characterized by a particularly harsh form of partisanship that not only sometimes blocks politicians from doing what is best for the country, it also has produced an electorate that is polarized and paralyzed. In this environment if one side is for something, the other side is often against it, regardless of its merit. We've witnessed the worsening gridlock for many years now.

It was unfortunate that former Vice President Al Gore's documentary film *An Inconvenient Truth* appeared on the scene in this sort of political climate. It more completely locked in the perception of global warming as a Democratic Party position and further intensified Republican Party opposition. Some Republicans in Congress have even tried to defund or otherwise prevent government agencies, such as the EPA and NASA, from conducting research or reporting on climate science-related topics. (Huffington Post - R. Zombeck)

Who is to blame? We all are for not only electing these politicians, but also for being sucked into the partisanship ourselves. We all need to get educated, throw off the partisan politics, and place our support behind politicians dedicated to saving the planet for human habitation.

Free-Market Capitalism

Free-market capitalism is an economic concept put forward by those who wish to create the appearance of working to minimize government "interference" in markets. Nearly all economists, politicians, and ordinary citizens agree that capitalism has been a major force behind the high-living standard we in the United States and the developed world enjoy. It is also widely accepted that government has an important role to insure fair play in commerce, and in other activities for the common good, such as national defense, law enforcement, critical infrastructure, environmental protection, and other essential social programs. In reality, "markets" don't exist without government. By definition, it is government's role to define the market rules, presumably for the common good of the country. When big money interests, like the fossil fuel industry, have the power to influence government, they attempt to set the "market" rules so they are "free" to do as they please. (Reich, 2015)

Pro fossil fuel politicians tend to deny climate science because they understand that solving the global warming problem, with its implications to regulate greenhouse gas emissions, government-supported renewable energy research and development, and laws to mandate energy efficiency and sustainable living practices are all actions to change the market at the expense of the fossil fuel industry. We've seen many instances, especially regarding environmental and health issues, when the common good justified government action. The many examples include leaded gasoline, unrestrained use of DDT and other pesticides, dumping of untreated sewage, PCBs, and other toxins, and sulfur dioxide emissions from coal fired power plants. These are just a few of the many examples where only a strong government representing the common good of the people could step in to dampen the unrestrained greed.

This doesn't mean government should run roughshod over the private sector. In fact, it is just the opposite. Government needs to support the many talented economists, scientists, and engineers in government, educational institutions, and the private sector, to define workable market rules and regulations that will allow the extraordinary power of capitalism and competitive markets to get the job done. Rather than hold back mankind, this is our opportunity to create a prosperous and sustainable economy built on renewable energy, instead of racing to the brink of extinction. Here again, who we vote for makes a big difference.

Religious Deniers

Throughout history, many conflicts have occurred as new discoveries clashed with religious ideology. Some religious followers continue to reject certain scientifically established truths that conflict with their religious dogma. The biblical story of creation is one of the most notable of these conflicts. To those who hold the literal word of their scripture as the absolute truth, certain scientific fields of study are to be refuted or avoided. Just as evolutionary science is often not accepted in these circles, the same is true of the Earth sciences. Many of these individuals feel their religion is under attack by science and attach the same distrust to anyone aligned with science such as scientific institutions, government agencies, and politicians who warn of human-caused global warming. Their distrust of science and lack of scientific knowledge make them easy recruits for global warming deniers who are eager to enlist their religious fervor in their obstruction.

Of course there are many religious leaders and their followers who *do* believe the climate science. They realize that we are clearly capable of doing much harm to our environment, as we have amply proven repeatedly. They see our greenhouse gas emissions as simply another example in the long history of human sin against the environment driven by greed and ignorance. As the chemist Jorge Mario Bergoglio, more famously known as Pope Francis, has said regarding climate change and how we have treated the Earth:

"Climate change is a global problem with grave implications: environmental, social, economic, political, and for the distribution of goods. It represents one of the principal challenges facing humanity in our day."

"If present trends continue, this century may well witness extraordinary climate change and an unprecedented destruction of ecosystems, with serious consequences for all of us."

"We have come to see ourselves as her lords and masters, entitled to plunder...."

Distraction, Indifference, and Complacency

In the United States, a strange change has taken place over the years regarding public attitudes toward global warming and climate change. Polls show that both the level of concern and the belief in a strong consensus have fluctuated considerably. What is causing this? (Climate Outreach Information Network)

Most of us live busy lives and have many priorities to juggle. We have families, jobs, churches, friends, sports, and so on. On a day-to-day basis, our focus is on immediate concerns such as getting to work on time or the kids off to school. We are exposed to all sorts of sensational and curious distractions in the media ranging from natural disasters, violence, scandals, and bad behaving athletes, pop stars, and politicians. We also have our smart devices to keep us plugged in to social media and whatever topic is our desire. With all of these distractions, where is the time for concern about global warming? It's like the old cliché: "We're busy rearranging the deck chairs on the Titanic." Here we are sailing along, going about our business, and enjoying the cruise, oblivious to the approaching danger. The Captain (our government in this analogy) who has been warned of the great danger is steering us to our end. The difference between our circumstances and the Titanic is that we passengers have not only been repeatedly warned of the danger, we have elected and reelected many leaders who have openly ignored the warnings. When our ship goes down, none of the many things that are distracting us now will be important.

If global warming is of such dire concern, why aren't we hearing about it constantly in the news media? Wouldn't it rise to the top of the "Breaking News" every day? To answer this we need to recognize that the "news media" is not focused on delivering what we most need to know. Collectively, the media is driven by viewership, readership, visits, and so on. They are about attracting attention. Climate science stories don't grab our attention like they once did. The subject has too much of the "same old story" feel to grab the headlines. It takes something dramatic to do that. Much of what has been in the media about global warming has been presented in ways that sensationalize the science or was so abbreviated that it couldn't help but tell an incomplete story and set inaccurate expectations. And it isn't just the media that has added to the confusion. Well-meaning celebrities, politicians, and some scientists have made exaggerated statements, perhaps in an attempt to elicit an immediate response. Even the weather news is peppered with flippant references questioning global warming in connection with isolated weather events. All of this has added to the skepticism.

The other truth about the news media is that they are not unbiased. The friends of the fossil fuels industry are advertisers, contributors, and owners of media outlets. They set the agenda for public service content on all the major media outlets, including public broadcasting. Occasionally stories and reports generated from the scientific community do break through. And when the media does report on them they are often cloaked in the denier's language designed to reinforce the notions of uncertainty and disagreement. Occasionally, news shows will attempt to show both sides of the issue by means of a debate with an individual representing the denier position on one side and a climate scientist on the other. This creates the impression that both sides possess equal weight, when in reality the scientific community overwhelmingly supports the position that global warming is unequivocally caused by human activity.

Collectively the global warming deniers have used the same methods that proved so effective in postponing action in the leaded gasoline and tobacco legal and political battles. The technique is to create doubt about the science, and especially to create the impression that the science is not settled. They have even created their own "experts" and "research" to refute the claims of climate scientists. And if any crack can be found in the climate science, such as overzealous scientists, any discredited research, or contradictory finding, they do their best to publicize it. (Forbes by Robert Wynne, 2014)

For public- and employee-relations reasons, many of the major fossil fuel companies have been forced to publically recognize the climate science in recent years. However, their efforts to fund denial and to postpone the transition to renewable energy go on. Although much of it is no longer direct, the funds continue to flow behind the scenes through industry organizations such as the American Petroleum Institute, the American Legislative Exchange Council, the Western States Petroleum association, political action committees, and individuals linked to the fossil fuel industry. (Guardian: Frumhoff and Oreskes , 2105)

THE PATH TO SUSTAINABILITY

"There is an urgent need to stop subsidizing the fossil fuel industry, dramatically reduce wasted energy, and significantly shift our power supplies from oil, coal, and natural gas to wind, solar, geothermal, and other renewable energy sources."- Bill McKibben - author, educator, environmentalist, and Co-founder of 350.org.

Global warming didn't happen overnight and it will take many decades to reverse the trend. This will require some significant changes to our energy sources, energy efficiency standards, and how we regulate greenhouse gas emissions. Some of the changes will be disruptive to established industries, professions, and the balance of power among nations. It will require international cooperation and coordination with governments of all major countries dedicated to this cause. Solving the crisis will require a renewable energy revolution that not only supplies abundant energy but also eliminates the pollution and the environmental destruction associated with fossil fuel extraction, production, and use. This is a positive scenario with the promise of higher living standards, a clean restored natural environment, healthier living conditions, and liberation from the grasp of wealthy energy-rich nations, cartels and companies. This is the only way to a sustainable future and a legacy to be proud of.

Regulating Methane, Nitrous Oxide, and Fluorinated Gases

Methane Regulation

Regulating methane emissions poses many challenges. These emissions occur so casually and from so many potential locations that it is hard to monitor and regulate. Some practices that are high emitting can be curtailed and where releases can be captured they should be. For example, many emissions from manure and landfills can be easily captured and burned to generate electricity, producing carbon dioxide emissions instead of the far more potent greenhouse gas methane.

Ironically, the initial transition to lower greenhouse gas emitting energy will include increased use of natural gas in place of coal because of the relatively low carbon dioxide emissions from natural gas. With the increased exploration, extraction, and use of methane, one can only imagine how many undetected methane seeps will occur in the far-flung gas fields, distribution systems, and storage facilities. Given the potent heat absorbing capacity of methane, it is possible that the benefits from the reduction in carbon dioxide emissions from switching from coal to natural gas may be completely offset or even made worse by increased methane emissions. This is one of those instances where we'll need to have some seriously effective regulation, monitoring, and enforcement.

Nitrous Oxide Regulation

Reigning in excessive use of nitrate fertilizer will require improved regulation over how and when it is applied and the enforcement to back it up. Because this will involve an education effort, licensing who can handle and apply nitrate fertilizers may be the best way to ensure compliance. Residential use will remain a difficult emission source to control. If public education efforts don't work, an outright ban on non-licensed nitrate fertilizer use might be the only solution. Significantly reducing nitrate releases from waste treatment and livestock production will also require stronger regulation and enforcement.

Although we already have regulations in place to reduce nitrous oxide emissions from many fuel combustion sources, more comprehensive regulations will be needed to further limit these unhealthy smog producing and potent greenhouse gas emissions. In the long term, the transition to electric transportation and non-combustion energy sources will dramatically reduce human-caused nitrous oxide emissions from fuel combustion.

Fluorinated Gas Regulation

Although long-standing international agreements are in place to eventually eliminate emissions of the ozone layer depleting fluorinated gases, the overall level of these gases has continued to increase. Some of this is because of the residue from refrigerants being released as old technology heating and cooling systems are removed and recycled. The other cause for the increase is that some of the gases now in use are not covered by the existing agreements. Because fluorinated gases are such efficient heat absorbers, additional agreements are needed to expand regulation to cover their greenhouse effects.

Geoengineering

Geoengineering involves manipulation of the environment to counteract or lessen the effects of global warming. Many ideas have been proposed ranging from seeding the sky with cloud producing chemicals and reflective particles, to spreading huge quantities of ground up rock into the oceans to offset acidification. Though some of these ideas may be feasible, there are typically many unknowns and most pose considerable opportunity for things to go wrong. And for some of the ideas the geopolitical issues are just as daunting. Imagine one country intentionally creating clouds that cover another country causing crop failure or other negative consequences.

Not all of the geoengineering ideas are radical, risky, or even politically difficult. Take some of the methods of capturing and storing carbon dioxide for example. It is hard to find much controversy in the idea of extracting it from the atmosphere by restoring natural forests or by somehow safely storing it underground.

Although there may be some geoengineering in our future, it shouldn't be attempted without first proving that all consequences are fully understood and acceptable for the common good, and never as an excuse to continue our greenhouse gas emitting ways.

Lifestyle Changes

The transition to renewable energy will affect where and how we live, and how we move around and interact with ours. Most urban areas in the United States have developed into low-density sprawl, where many of us live away from urban centers and are primarily dependent on personal automobiles for transportation. If you put a high value on energy efficiency, this is not the ideal arrangement. And as many of us have learned through personal experience, spending so much time commuting isn't the best use of our precious time.

Future urban development will most certainly reflect energy efficiency and environmental concerns, as well as higher life quality goals. Already we're seeing new urban development where housing, retail, and work centers are being built along mass transit corridors, with serious attention given to integrating transportation networks to get us out of our cars so we can spend more time enjoying life and being productive. We're also seeing revitalization of urban centers with jobs, retail, housing, and destinations such as sports, entertainment, and cultural attractions locating there. People are increasingly valuing the vibrancy and convenience of city living. As these trends take hold, urban areas of the future will be far more energy efficient, pedestrian and bicycle friendly, and evermore livable.

Living a sustainable lifestyle also means being more attentive to the environment, conservation, reuse, and energy use in the choices we make about most everything, including the food we consume and even the clothing we wear. The motivation for these decisions won't merely be limited to the desire to do what is best for the planet. As the world population and living standards rise, we'll experience higher costs for many increasingly scarce natural resources and commodities. And as greenhouse gas emissions are increasingly regulated and their externalized costs are priced into products, there will be financial reasons to conserve and to avoid products with high global warming impact. The more of us thinking about sustainability the better it will be for everyone.

Energy Efficiency

The most immediate impact we can have on reducing greenhouse gas emissions is to be more efficient in our use of energy. We can do many things that are relatively easy and non-disruptive. Simply switching to efficient lighting and appliances, sealing up our windows, doors, and air ducts to prevent loss of conditioned air, and being more sensitive to the use of lighting and heating and cooling systems can make a meaningful difference in our energy consumption. Larger investments, such as installing energy-efficient windows, improved insulation, and high-efficiency heating and cooling systems can make even more of a difference.

Most of us can make big improvements in energy use and greenhouse gas emissions from transportation. Driving our cars less, walking or bicycling more, using mass transit, telecommuting, and buying locally produced and sourced goods are among the work and lifestyle changes we can all make. Although exchanging our fossil-fuel hungry vehicles for more efficient and lower greenhouse gas emitting ones may involve some investment, adjustments, and compromises, we will realize both financial benefits as well as the personal satisfaction of doing what is best for the planet.

Transitioning from Fossil Fuels

Our current global economy depends on fossil fuels. Although abundant renewable sources of energy exist that can be used to replace these fuels, it will take serious investments and decades to fully develop them and to transition our energy infrastructure to fully utilize them.

One of the arguments used in fighting action to curtail fossil fuel use is the claim that transitioning to renewable energy will slow or even cripple our economy. The fallacy in this reasoning is twofold. Many studies have clearly shown that in the long run renewable energy will be far less costly. Of course this is not good news to the fossil fuel industry, as they will see their wealth considerably diminished. The second point is that the many forms of pollution from fossil fuels and their global warming effects are already producing strong headwinds for our economy, and they will become catastrophic if we allow global warming to continue unrestrained. (International Monetary Fund Fossil Fuel Cost) (Romm, 2015) Far from crippling our economy, transitioning to renewable energy will save our economy and produce a much more equitably shared wealth.

We can tackle carbon pricing in numerous ways. Gradually increasing fees up to the cost of removing carbon from the environment or to equal a calculated "social cost" could be directly applied to each type of fossil fuel based on carbon content. These gradually imposed fees could also be used to fund credits to those using renewable energy. *Cap-and-trade* is a market-based approach designed to create financial incentives and to reward innovation. It operates by phasing in a cap on the target emission and by providing the emitters flexibility in how they comply, including financial penalties for not complying. We know cap-and-trade programs work because we used this approach following the 1990 amendment to the Clean Air Act to reduce the acid rain producing sulfur dioxide and nitrogen oxide emissions from coal-fired electrical power plant. This program was successful, surpassing the emission reduction goals and at much lower than the predicted cost. Another powerful government tool is to regulate greenhouse emissions as "pollution" through the Environmental Protection Agency (EPA). Any of these approaches will require a clear majority of elected officials at all levels who recognize that human-caused global warming is an important issue and who have the public support to take corrective action.

Converting to a renewable energy economy will require considerable new energy infrastructure as well as replacement of equipment at the point of consumption, such as personal and commercial vehicles and heating systems. Here is a critical role for government to step in to make the necessary incentives available. And by applying readily available energy efficiency measures, we can easily avoid negative economic consequences during the transition. There is every reason to believe that the investment in renewable energy development and infrastructure will provide a significant economic stimulus. Finally, when all costs are considered, the transition to sustainable renewable energy will be far less costly than staying the course using dirty fossil fuels. Why am I so certain? What could be more costly than making the planet uninhabitable?

Winners and Losers

Although the transition to renewable energy will be a big win for mankind generally, there will be differences in terms of the resulting wealth and prosperity among individuals, industries, regions, and nations. We've already addressed the fossil fuel investors who stand to see significantly lower valuations on their investments. This also includes some fossil fuel rich countries that will experience big changes in world standing and wealth.

Some fossil fuel interests may benefit, at least temporarily, during the transition to curb greenhouse emissions. As already mentioned, the natural gas segment is a great example of this. Because burning natural gas produces far less carbon dioxide per unit of heat than does burning coal, carbon cap-and-trade or carbon taxing will create incentives to switch from coal to natural gas. We are already seeing strong signs of this shift in the United States, although the current rush to natural gas for power plants is more a result of the low cost of natural gas than any desire to reduce carbon dioxide emission. It does clearly show how market pricing can be used to motivate large-scale transitions to lower carbon-emitting fuels. As fossil fuels are gradually phased out, generally low profit margins will also place some fossil fuel interests at competitive advantage to others. Hard to extract and refine tar sands oil for example, will not be competitive with the "sweet crude" easily flowing out of Saudi Arabia.

There will be winners who are involved in energy efficiency and with carbon dioxide emission mitigation. In a cap-and-trade situation, one of the ways a carbon dioxide emitter can earn credit offsets for their emissions is by removing carbon dioxide from the environment in some fashion, such as by increasing photosynthesis. Businesses and governments involved in helping companies earn these credits by planting trees or restoring habitat fall into this category. Companies involved in the technology to capture carbon dioxide at stationary fossil fuel power plants may also be winners.

There will be many winners among the innovators and investors who pioneer and develop the new energy sources and infrastructure of the renewable energy revolution. Their workers will also be winners, as will the communities where they live and work. Countries like the United States will no doubt share in this activity and benefit greatly. However, our failure to aggressively pursue a transition to renewable energy over the past few decades has already provided opportunities for other countries such as Germany and China to get a leg up on us. The longer we wait, the farther we fall behind. (PEW - Energy Race)

Beyond those who own fossil fuel assets, many others will be potential losers in the renewable energy transition. People working in fossil fuel and supporting industries will be directly affected. Though the transition will be gradual and allow for natural attrition of workers such as through retirement, some will face the prospect of skill obsolescence and changed employment prospects. And it doesn't stop with just those directly working in fossil fuels. Any industry that will see significant technology change will also be touched. For example, you can count on the transportation, utility, and heating and air conditioning industries to see big changes as new technologies are introduced requiring new skills.

Disruptive technologies always force change. We have the opportunity to plan for retraining, retooling, and other support so that we avoid much of the income insecurity, unemployment, and dislocation that often come from such transitions. Regardless of how well intended and well executed these efforts are, it is important to remember that people don't like forced career change or employment uncertainty. It will not be easy for a U.S. Senator from Oklahoma to convince the oil and gas industry workers of that state that they should vote to lose their jobs and trust that they'll find new careers in renewable energy.

Given the gradual nature of the renewable energy transition, the long term lower cost of renewable energy and the eventual placement of so much new infrastructure, the net effect on our ongoing economy will be a positive one. (UCS Renewable Benefits) Other nations that have taken a more aggressive stance in transitioning to renewable energy have already shown this to be true. This is especially obvious when you consider the devastation to our economy from not doing anything to curb global warming.

Engineering the Transition to Renewable Energy

We experience energy in many ways: from the heat of the sun and strength of the wind, to the force of ocean waves and currents, and at times from heat welling up from deep in the Earth. In fact, we are surrounded by enormous amounts of unharnessed energy. Consider this: The solar energy falling on the Earth every hour easily exceeds mankind's energy needs for an entire year! (Stephen Cass, 2009) (SOLAR FAQs, 2006) And this doesn't include the energy we could produce from wind, nuclear, geothermal, and tidal sources. The renewable energy transition is about making use of this incredible abundance.

A useful approach to studying the transition to renewable energy is to focus on *where* the disruptions in the source of energy and the equipment required to produce or use the energy are. For example, we get much of our energy through the electrical grid. As electric utility consumers, we are essentially indifferent to how the electricity is generated. Our devices work the same regardless of the source. When switching to renewable energy, most of the disruption in the electrical grid will be at power plants and with the fuels they use. Businesses, governments, and individuals will face potential disruptions where they directly burn fossil fuels, such as for heating, transportation, and cooking.

Although some situations will exist where renewable fuels will not require significant change to the equipment consuming the fuel, such as the use of some biofuels, most will likely require complete replacement. All of this equipment has a natural replacement age anyway, so a reasonable transition would be a phased approach in which the old equipment is gradually discontinued and replaced with the new. We've experienced this kind of planned transition before. Consider how much cleaner, more efficient, safer, and more reliable our automobiles are today than they were forty years ago. And all of this was accomplished without upending the economy while saving countless lives, improving the environment, and providing all of us with a better transportation experience. We should expect that the gradual transition to renewable energy would be just as beneficial.

Carbon Capture and Storage

Technologies and practices that capture and store carbon may play a role in reducing atmospheric carbon dioxide. The most basic of these approaches is simply to increase the amount of photosynthesis by growing more plants. As already mentioned, carbon dioxide emitters can use habitat and forest restoration as offsets to achieve net neutral carbon status. To the extent that we can rely on replanting that simulates natural regrowth, mankind and many animal and plant species will undoubtedly benefit greatly from such restorative efforts.

Carbon dioxide scrubbing refers to technologies that capture carbon dioxide from combustion exhaust streams, such as at coal power plants. This equipment is expensive and also requires a place to permanently store the captured carbon dioxide. Injecting the carbon dioxide into underground chambers or deep into the oceans, where it will presumably not escape, is the common suggestion. Though the technology to capture the carbon has already proven to be feasible, it is prohibitively expensive and dramatically reduces the net energy produced from the power plants. Finding adequate and secure storage for the captured carbon dioxide may be its practical limitation.

A similar approach is to separate carbon from hydrocarbon fuel such as methane (CH_4) prior to combustion. The captured carbon (C) solid is then safely stored in a stable environment and the remaining hydrogen (H_4) used as a transportable chemical or combustible fuel. Although this approach is feasible, like carbon dioxide scrubbing it is expensive and because the process requires so much energy it only makes sense if powered by abundant renewable energy.

Electricity from Heat

Much of the world's electric power is generated using heat to drive steam-powered electric generators. The heat can come from many sources.

Concentrated solar uses mirrors to focus and concentrate the sun shining on a solar field to generate heat. Although it has the disadvantage of being available only when the sun is shining, it can be effectively used to produce energy well beyond sunny periods if some of the energy produced is also stored for later use. Although some concentrated solar installations have had some negative effects on sensitive wildlife, proper consideration to site location, design, and operating procedures to avoid bird kills from exposure to the concentrated heat have been helpful in reducing these effects. (Daily Mail - Sarah Griffiths)

Around the world, countries are showing a reluctance to continue to expand the use of nuclear energy. Not only have some of these facilities proven to be unsafe, it has also been difficult to safely handle and store the nuclear waste. The cost of nuclear plants and their operation has also been prohibitively high. To significantly impact our future energy needs with nuclear, we will likely need to develop some different reactor technologies that address these issues and embark on an unprecedented and probably unrealistic building schedule. The future doesn't look so good for nuclear. (BBC - Richard Anderson, 2015)

Geothermal is especially attractive in locations with relatively easy access to tremendous amounts of underground heat such as many of the Western states where geothermal power has long been used to produce large amounts of electricity.

All of these carbon-neutral heat sources are already in use and available to replace carbon dioxide emitting fuels such as coal, oil, and natural gas. We should expect increased use of concentrated solar and geothermal, and declining use of nuclear.

Electricity from Wind and Water

Wind, water currents, and falling water all represent some of the oldest and most familiar forms of energy to run mills, pump water, propel water vessels, and generate electricity. Though these sources often feel like they are free, this is far from true. For example, large-scale hydroelectric power invariably relies on dams that typically flood land that would otherwise contain carbon-storing plants. Dams not only seriously disrupt fish and other plant and animal life, they often produce significant methane emissions from decaying vegetation, displace people, and bury historical and sacred sites. And when you consider the relatively short life of many dams due to the accumulation of silt, and the carbon dioxide emissions from the concrete used to build them, they often aren't the best choice. A related technology for storing energy from renewable sources is the practice of impounding water by pumping it uphill when there is an abundance of the renewable energy. The water is later released to flow downhill to generate electricity from the force of the falling water. Small-scale hydroelectric is a promising source of energy. This refers to the installation of small generators that are driven by existing water flows such as in natural fast-flowing streams or in industrial environments such as water treatment facilities and factories. (Small Hydro International Gateway) The biggest threat to hydroelectric power is the diminished supply of water in many areas where climate change will produce lower levels of precipitation.

Many U.S. states and nations across the globe have more than enough potential wind power to provide 100% of their energy needs. Wind turbine-generated electricity does come with some environmental consequences. The early generations of wind turbines were particularly destructive to birds. Newer turbines, which are positioned much higher off the ground and turn more slowly, are typically placed out of frequently used bird flight paths. Some are even utilizing radar capable of detecting the presence of a bird and are able to quickly shut down to avoid collisions. Wind turbines also scar the landscape and damage wildlife habitat, as anyone who has ever observed a large-scale wind farm can attest. On the positive side, wind turbines take up relatively little area and the surfaces beneath them are still available for other purposes such as agriculture or aquaculture. Wind is proving to be a reliable and relatively low cost source of power. We can expect that it will likely be one of the primary sources of energy in our future.

Ocean wave, current, and tidal generators may also prove to be viable sources for electric power generation in areas with particularly strong and consistent ocean wave or tidal forces, though probably not without imposing some costs to ocean wildlife.

Electricity from Photons

The photovoltaic effect, discovered in 1839 by nineteen-year-old Frenchman Edmund Becquerel, directly produces electricity from sunlight. Photovoltaic "solar panels" use materials that release electrons as they absorb photons of light to produce electricity. The electric current produced can be directly used or it can be converted to alternating current through a device called an *inverter* for use with a conventional electrical power grid. Solar panel technology advancements have dramatically improved the productivity of these panels in recent years while also substantially reducing the cost. So much so that even without government incentives, most solar panel installations already enjoy a reasonably short cost recovery period. And in especially sunny locals, most are capable of supplying far more than is needed for much of the day. The excess can be stored in various ways for use when the sun isn't shining or it can be supplied to the local utility. All is not resolved in many locations with the local utilities however. Not only do they need to be capable of managing this flow of excess solar panel electricity into their systems, they also need to have their business models adjusted to complement private solar panel use as well as energy conservation. This isn't always an easy transition for these businesses accustomed to operating under a government-supported energy monopoly. Large utility scale solar panel installations can also have some negative effects on wildlife through habitat destruction and to birds that mistake their reflective surfaces for water. This "lake-effect" can be fatal to water birds harmed while landing and especially to those species incapable of taking off out of water. Proper site location and design considerations to avoid attracting water birds are essential to avoiding these unacceptable consequences.

As solar panel and on-site power storage technology continues to improve, more home and commercial locations will be in the position to essentially "drop off" the utility grid much of the time, if not all together. For many the cost of solar power is already close to par, or lower, than the cost of fossil fuel and nuclear sources. As the price of solar-generated electricity continues to fall, and it most certainly will, market forces will drive another nail in the fossil fuel coffin. Putting a price on carbon dioxide emissions will certainly hasten this transition.

Energy from Biofuels

Biofuels are produced from organic matter. They are described as "carbon neutral" because the carbon dioxide produced from burning them is offset by the carbon dioxide removed from the atmosphere when their source plants were only recently grown. The result over this relatively short period from plant growth to combustion is no net change in atmospheric carbon dioxide. Burning fossil fuels, on the other hand, frees carbon long removed from the atmosphere causing a net increase in atmospheric carbon dioxide.

Ethanol, a biofuel commonly blended with gasoline here in the United States, is primarily produced from corn. One of the major problems with this plant source is that it competes with corn produced for food, raising the price of this commodity that so many in the world rely on. Another issue with ethanol is the high level of energy required to produce it, typically using fossil fuel sources. Our use of ethanol has so far been more about the power of the agriculture lobby to increase demand for corn than any effort to reduce carbon dioxide emissions. In other countries, such as Brazil, sugar cane is the primary ethanol source crop.

As previously mentioned, methane gas is produced from the decomposition of organic material. When this gas is contained at biomass waste treatment plants and landfills for use as fuel it is also considered a biofuel.

Vegetable oils can also be used as biofuels. Production of vegetable oil often competes with food crops and has been responsible for some significant conversions of natural habitat to plantations. Oil produced from algae is an example of a biofuel that isn't from a human food crop. Algae oil can be used as a replacement for fossil crude oil, meaning it can be refined to produce the same products we commonly associate with petroleum such as gasoline, diesel, and jet fuel. This is logical when you consider that crude oil is formed from ancient deposits of phytoplankton consisting primarily of algae. For mobile energy needs, such as jet aircraft, where high-energy dense combustible fuel is essential, algae-derived fuel is a potential replacement. Many researchers are studying algae to identify what strains work best and how to scale it for mass production. The algae needs sunlight, carbon dioxide, adequate water, essential plant nutrients, and, of course, an environment that optimizes production and facilitates harvesting. (U.S. Department of Energy: ALGAE-TO-FUEL) Methods for producing carbon-neutral combustible fuels also exist that do not rely on photosynthesis and organic matter. There are many possibilities for producing energy-rich portable carbon-neutral fuels that are waiting in the wings for a cost to be applied to carbon dioxide emissions so that they may be more competitive.

Perhaps the greatest negative about combustible carbon-neutral fuels is that, like fossil fuels, the incomplete combustion that occurs when they are burned generates pollution. In addition, combustion engines are relatively inefficient because they lose so much energy in the form of waste heat. Add in the high cost of producing and maintaining these complex engines and their associated power transmission systems, and future reliance on combustion fuels and engines will most certainly diminish in favor of electric power.

Power from Combustion Engines

Fossil fuel burning combustion engines are a major source of greenhouse gas emissions. They include the piston, turbine, and jet engines used in passenger cars, trucks, ships at sea, locomotives, and aircraft. We also use them to power all sorts of other portable and stationary equipment. The transition to renewable energy will require that we eliminate the carbon dioxide and other harmful emissions from these engines. This isn't easily done in a short period of time and will require a gradual transition. Some uses for combustion engines will be more difficult to transition than others. As previously mentioned, jet aircraft, the most carbon dioxide emission-intensive form of transportation may always need carbon-dioxide emitting fuel. Large trucks and ocean-going ships may also need to rely on the energy rich liquid fuels used in combustion engines. Achieving net zero carbon dioxide emissions for any combustion-engine-powered form of transportation will require a carbon-neutral fuel or offsets through some form of carbon storage.

Many of the devices we power with combustion engines can be powered from electricity, either connected directly to the grid, or indirectly through batteries charged from the grid or other sources, such as solar panels. It is the mobile power applications where the switch to electricity provided from batteries is most challenging.

Batteries and Other Forms of Energy Storage

Chemical batteries are everywhere in our daily experience. In the renewable energy transition, rechargeable chemical batteries will play a big part in storing energy for use in both stationary and mobile applications. Chemical batteries have long provided back-up electric power for computer and other critical electric applications. They are also commonly used to store energy generated from solar cells for use when the sun isn't shinning.

In spite of recent advancements in the technology, chemical batteries continue to have a much lower energy density than do fossil fuels. For example, the amount of space and weight required for the gasoline or diesel fuel to propel an automobile hundreds of miles is relatively little when compared to the comparable space and weight required for chemical batteries to accomplish the same task. The other disadvantages of chemical battery power for automobiles compared to gasoline or diesel is that it takes much more time to replace or recharge batteries than it does to merely refill the fuel tank. Chemical batteries also must be periodically replaced because they gradually lose their ability to take a charge.

Although battery-powered automobile motors are much more efficient than combustion engines, the added bulk and weight and the long recharge time of chemical batteries more than offset the efficiency advantage, making the electric auto less practical where long driving distances are required. But for relatively short distance driving, such as the typical daily distance in the United States, electric vehicles are practical and affordable. The size and weight of the required battery for short-range driving isn't prohibitive and the recharge time can easily be accomplished during the time when the vehicle is parked.

Plug-in-gas-electric vehicles are designed for situations when we need a longer driving range. These vehicles run on battery power then switch to combustion engine power and convenient fuel refills when their battery is discharged. Although most of the electric and plug-in-gas-electric vehicles now on the market are relatively small and don't project the image of performance and prestige, many are environmentally friendly and economical. As the several high-end electric passenger vehicles now on the market have demonstrated, it is possible to build electric vehicles that have great performance, style, safety, and comfort, and still have a range in excess of several hundred miles. As chemical battery technology continues to evolve producing lighter, smaller, longer-lasting, and faster-charging batteries, and as electric vehicle production volumes grow to achieve economies of scale, the price of electric vehicles will fall and they will become increasingly more desirable.

The hydrogen-powered fuel cell is a source of portable electricity that may play a part in our future. It has been used for many decades in spacecraft. Because hydrogen combines so easily with other elements it is primarily found on Earth in molecular form such as water (H_2O) and methane (CH_4). Producing pure hydrogen requires separating it from such molecules. This processing consumes a lot of energy and because hydrogen gas is light and ignites easily, it must also be carefully contained. This makes producing and handling it both expensive and rather dangerous. The primary emission from a hydrogen fuel cell is water and some waste heat. Hydrogen fuel cell test vehicles have already been in everyday use in California and elsewhere for several years. Refueling them is much like a gasoline-powered car through a nozzle and hose to fill the high-pressure tank. The limiting factors are the lack of refilling stations, the high cost of producing hydrogen, its relatively low energy density, the high price of hydrogen fuel cells, and difficulties associated with safely storing and handling hydrogen. Although California is moving forward to create a hydrogen-refueling infrastructure as the sale of hydrogen-powered vehicles begins there, these are difficult hurdles to overcome. Creating a carbon-neutral hydrogen ecosystem will also require that the hydrogen be produced using non-carbon or carbon-neutral sources and distribution methods.

There are other ways to store energy besides combustible and chemical sources. Heat can be stored, such as in a molten salt solution to later drive an electric generator. Some concentrated solar power plants use this technique. Another approach is to lift something to store energy expressed as the force of gravity acting upon it. Pumping water uphill was previously described as an example of this approach to later generate electricity when the water is released to flow downhill to drive a hydroelectric generator. Compression of gas in a closed container can be used to later provide energy when it is released. Or the energy of a mass in motion, such as a spinning flywheel, can be used to store energy. Many of these methods are already used in industrial and utility environments to store energy and to stabilize otherwise variable electricity sources. They will likely become increasingly important components in utility, residential, and commercial energy installations.

Enough to Go Around

Population, living standards, and energy efficiency determine energy consumption. Given expected global population growth and rising living standards across the planet through the end of this century, worldwide energy consumption will be a multiple of present levels, likely more than triple. Will we have enough energy to satisfy the demand once we wean ourselves from fossil fuels? The short answer is yes, more than enough, and with plenty to spare! To understand the big picture and what changes we can expect, we need to look at *where* the energy will come from, how it will be delivered to us, and the value we will need to put on energy efficiency. (IPCC AR5, 2014) (WWF - Renewable Energy)

Of all the potential renewable energy sources, the energy we receive directly from the sun dwarfs all the others. The amount of solar energy we can reasonably expect to make use of, given how much of the Earth's surface is covered by land and that has a desirable sun exposure, is truly enormous and far exceeds our projected needs many times over. This abundance can be used to produce electricity through photovoltaic processes, concentrated solar, and to produce transportable combustion or chemical fuels. (SOLAR FAQs, 2006)

The tropic zone, between the Tropic of Capricorn in the southern hemisphere and the Tropic of Cancer in the northern hemisphere, has the most advantageous average solar exposure. The Earth's temperate zones that include much of North America, Europe, and large parts of Asia, Australia, and South America, have a slightly less desirable solar exposure, though still relatively productive. Closer to the poles the solar exposure is much less intense. As solar begins to play an ever bigger part in the energy picture, the balance of power and energy interdependence among nations will most certainly change. We are again reminded of the importance of energy in defining the world order. The United States is fortunate because we are well positioned both geographically and technically to take advantage of the renewable energy transition. (SOLAR FAQs, 2006)

We will, by necessity, become much more focused on energy efficiency. This will extend through our personal habits, our transportation, how our energy-using devices are engineered, and the design of our energy infrastructure. Our electrical grid, for example, will be different. Today we lose a considerable amount of energy to waste heat from line resistance as we send electricity through transmission lines over great distances. Placing the renewable power generation sources closer to the point of consumption will significantly reduce this loss. Other changes we can expect include:

- Photovoltaic panels and electric storage will become standard components in our residential, commercial, and public structures.
- Combustion fuels for cooking, heating, and transportation will increasingly give way to electrification.
- Electric vehicles and charging stations will be everywhere.
- Community and cooperative renewable energy facilities will be commonplace.
- Far-flung electric power generation plants away from population centers such as large geothermal, wind farm, concentrated solar, nuclear, and hydroelectric dams may still exist to provide base load power, but much of

their power may also be used to produce transportable fuels instead of electricity.

- Shipment of goods and transportation requiring energy dense combustion fuels will be more expensive.
- High-speed electrified rail will serve all of our major intercity corridors.
- More use of mass transit and of car sharing and rental for travel out of the immediate area.

During the transition to renewable energy, the role of utility and giant energy companies will undoubtedly change. We will still need large companies capable of developing and building the renewable energy plants, production facilities, and the distribution systems. These companies will be valued on their ability to produce and deliver renewable energy rather than on their fossil fuel reserves and their political clout to block the transition to renewable energy. Power utilities will also undergo a transition from publically supported energy monopolies to integrators of the various energy sources and managers of energy systems.

Although the transition to 100% renewable energy in the U.S. will require some significant changes in our energy production and distribution infrastructure, it doesn't require new breakthroughs and is achievable using current technology within several decades. (Jacobson, 2015) (Mark Z. Jacobson, 2015) This isn't to say that there aren't areas of technology where breakthroughs would be helpful, because there are. The point is that waiting isn't necessary or even helpful. The price for wind and solar energy is already competitive and much of the world is ready to ramp up. A world economy built on renewable energy will be far better for mankind and the planet. The sooner we get started the better.

A MORAL OBLIGATION

In *THE FORCES OF DENIAL* section, I described externalized costs from greenhouse gas emissions that allow those who own the various entities responsible for these emissions to profit enormously at the expense of the rest of the world's present and future population. I portrayed those responsible for the tremendous harm caused to others and the truly enormous transfer of wealth from the many to the few as being selfish and immoral. Is this too harsh? Let's consider who these people are. It isn't just the energy-rich countries and companies producing and selling these products and their allies in government. All of us who directly or indirectly use these products and contribute to the emissions are also complicit. Isn't our unwillingness to put things right and to continue pushing these externalized costs off onto others and future generations also selfish and immoral?

Knowing right from wrong defines the morality line. My hope is that you now understand the truly catastrophic consequences of global warming, what is causing it, who is denying it, what motivates them to do so, and what mankind must to do to solve this problem. Which side of the morality line will you be on?

Standing up for a transition to renewable energy may not be easy. You may find considerable resistance from family, friends, neighbors, and perhaps members of the political, social, and faith groups you belong to. And those who own fossil fuel assets will continue to fight to protect their wealth. The politicians, lobbyists, and big money public relations firms they employ will do all they can to continue their immoral denial of human-caused global warming from greenhouse gas emissions. Those opposing you will include many who are merely uninformed and many others who have been intentionally misled. Will you have the strength and patience to calmly engage with them to spread the wisdom? You will also find some opposing you who are simply selfish, who don't care about others, either alive now or in the future. Will you have the fortitude to push through this opposition to financially back and elect leaders who will support public policies to transition to renewable energy and a future saved from the devastating effects of global warming?

We in the United States are truly at a most critical crossroad. We can continue to deny the global warming reality and follow the road to catastrophic environmental disaster and unimaginable human suffering. Or, we can take the road to a sustainable future. If we act soon and with enough conviction, we can join with other nations to avoid some of the worst and begin the process of establishing a far better life for all mankind. Just imagine a world free of the pollution and environmental destruction from consuming fossil fuels. And because the United States will be energy rich with abundant renewable energy sources, our military will no longer need to be focused on protecting energy supplies from across the globe. Wealth from energy will no longer be concentrated in the hands of a few fossil-fuel rich nations, corporations, and individuals. Instead, energy ownership will be more local, belonging collectively to the citizens who reside nearby, and individually to those who invest in capturing the energy that literally falls freely upon them.

Our descendants will look back on these past few decades in wonder. Regardless of what action we take now, they will likely find our past denial of human-caused global warming as unforgivable. What is uncertain is just how much they will suffer. If we fail to act soon, they will most certainly suffer far more or perhaps fail to exist at all.

"Sustainable Development *is development that meets the needs of the present without compromising the ability of future generations to meet their own needs."* - International Institute for Sustainable Development (IISD)

The sustainable way is the moral way.

SUGGESTED READING

- *Dire Predictions: Understanding Climate Change 2nd Edition* by Michael E. Mann and Lee R. Kump, 2015 DK Penguin Random House
- *This Changes Everything: Capitalism vs. The Climate* by Naomi Klein, 2014 - Simon & Schuster
- *Climate Change: What Everyone Needs to Know* by Joseph Romm, 2015 - Oxford University Press
- *The Sixth Extinction* by Elizabeth Kolbert, 2014 - Henry Holt and Company, LLC
- *Tropic of Chaos: Climate Change and the New Geography of Violence* by Christian Parenti, 2012 - Nation Books
- *Hot, Flat, and Crowded: Why We Need a Green Revolution—and How It Can Renew America* by Thomas L. Freidman, 2008 - Farrar, Straus and Giroux
- *Oil and Honey* by Bill McKibben, 2013 - Henry Holt and Company, LLC
- *The Age of Sustainable Development* by Jeffrey D. Sachs, 2015 - Columbia University Press
- *The Great Transition: Shifting from Fossil Fuels to Solar and Wind Energy* by Lester R. Brown, 2015 - W. W. Norton and Company, Inc.
- *Harness the Sun: America's Quest for a Solar-Powered Future* by Philip Warburg, 2015 - Beacon Press
- *Unstoppable: Harnessing Science to Change the World* by Bill Nye, 2015 - St. Martin's Press
- *Atmosphere of Hope: Searching for Solutions to the Climate Crisis* by Tim Flannery, 2015 - Atlantic Monthly Press
- *Climate Cover-Up: The Crusade to Deny Global Warming* by James Hoggan, 2009 - Greystone Books - D&M Publishers, Inc.
- *The Hot Topic: What We Can Do About Global Warming* by Gabrielle Walker and David King, 2008 - Harcourt, Inc.
- *Our Choice: A Plan to Solve the Climate Crisis* by Al Gore, 2009 - Rodale Books

BIBLIOGRAPHY

350.ORG, D. t. (n.d.). Retrieved October 2015, from http://math.350.org/

BBC - Richard Anderson. (2015, February 27). *Nuclear power: Energy for the future or relic of the past?* Retrieved March 10, 2016, from http://www.bbc.com/news/business-30919045

Carbon Tracker Unburnable Carbon. (n.d.). *Wasted capital and Stranded Assets.* Retrieved February 16, 2016, from http://www.carbontracker.org/report/wasted-capital-and-stranded-assets/

Center for Climate and Energy Solutions. (n.d.). Retrieved October 2015, from http://www.c2es.org/facts-figures/main-ghgs

Climate Outreach Information Network. (n.d.). *Public perceptions of climate change.* Retrieved October 2015, from http://talkingclimate.org/guides/public-perceptions-of-climate-change/

Daily Mail - Sarah Griffiths. (n.d.). *Solar farm sets 130 birds on FIRE: Extreme glow of power plant ignites creatures mid-air during tests.* Retrieved March 10, 2016, from http://www.natureworldnews.com/articles/12918/20150223/solar-farm-set-hundreds-birds-ablaze.htm

EarthSky. (n.d.). *How much do oceans add to world's oxygen?* Retrieved February 16, 2016, from http://earthsky.org/earth/how-much-do-oceans-add-to-worlds-oxygen

Ellie Zolfagharifard for Daily Mail. (n.d.). *Massive ice shelf in Antarctica could break up.* Retrieved October 2015, from http://www.dailymail.co.uk/sciencetech/article-3079089/Massive-ice-shelf-Antarctica-break-without-warning-say-scientists-cause-havoc-coastlines.html

Emissions from Cement Industry. (n.d.). *Columbia University, Earth Institute, Climate.* Retrieved from http://blogs.ei.columbia.edu/2012/05/09/emissions-from-the-cement-industry/

EPA U.S. Climate Models, U. E. (n.d.). *US Environmental Protection Agency.* Retrieved October 2015, from http://www3.epa.gov/climatechange/science/future.html#Temperature

Forbes by Robert Wynne. (2014, June 26). *Forbes, The Public Relations Debate About Global Warming Heats Up.* Retrieved October 2015, from http://www.forbes.com/sites/robertwynne/2014/06/26/the-public-relations-debate-about-global-warming-heats-up/

Forbes.com -Susan Adams - Why Selfish Leaders Triumph. (2011, September 29). *Why Selfish Leaders Triumph.* Retrieved January 27, 2016, from Forbes: http://www.forbes.com/sites/susanadams/2011/09/29/why-selfish-leaders-triumph/#71ae0b452442

Guardian: Frumhoff and Oreskes . (2105, March 25). *Fossil fuel firms are still bankrolling climate denial lobby groups.* Retrieved March 29, 2016, from http://www.theguardian.com/environment/2015/mar/25/fossil-fuel-firms-are-still-bankrolling-climate-denial-lobby-groups

Haidt, J. (2012). *The Righteous Mind: Why Good People Are Divided by Politics and Religion.*

Huffington Post - R. Zombeck. (n.d.). *GOP Works to Defund Studies So They Can Deny Climate Change.* Retrieved March 10, 2016, from Huffington Post: http://www.huffingtonpost.com/richard-zombeck/gop-works-to-defund-studi_b_7279786.html

IISD. (n.d.). *International Institute for Sustainable Development.* Retrieved from https://www.iisd.org/sd/

International Monetary Fund Fossil Fuel Cost. (n.d.). *IMF: 'True cost' of fossil fuels .* Retrieved March 10, 2016, from Public Radio International: http://www.pri.org/stories/2015-06-07/imf-true-cost-fossil-fuels-53-trillion-year

IPCC AR5. (2014). *Fifth Accessment Report.* Intergovernmental Panel on Climate Change.

IPCC. (n.d.). *Intergovernmental Panel on Climate Change.* Retrieved October 2015, from http://www.ipcc.ch/

Jacobson. (2015, May 27). *100% clean and renewable roadmaps for 50 states.* Retrieved November 4, 2015, from http://web.stanford.edu/group/efmh/jacobson/Articles/I/USStatesWWS.pdf

Katie Jennings, D. G. (2015, October 23). *How Exxon went from leader, LA Times.* Retrieved October 25, 2015, from http://graphics.latimes.com/exxon-research/

Mann and Kump, M. E. (2015). *Dire Predictions 2nd Edition.* DK Penguin Random House.

Mark Z. Jacobson. (2015, June 8). *Stanford engineers develop state-by-state plan to convert U.S. to 100% clean, renewable energy by 2050.* Retrieved November 4, 2015, from Stanford News Service: https://news.stanford.edu/pr/2015/pr-50states-renewable-energy-060815.html

NASA Earth Observatory, G. W. (n.d.). *How Much Will Earth Warm.* Retrieved October 2015, from http://earthobservatory.nasa.gov/Features/GlobalWarming/page5.php

NASA Facts about Aerosols. (n.d.). *Atmospheric Aerosols.* Retrieved February 16, 2016, from https://www.nasa.gov/centers/langley/news/factsheets/Aerosols.html

NASA Public Perception of Climate Change. (n.d.). *National Aeronautics an Space Adminisatration, Goddard Institute of Space Studies.* Retrieved October 2015, from http://www.giss.nasa.gov/research/briefs/hansen_17/

Natural Resource Defense Council - James Inhofe. (n.d.). *NRDC Action Fund.* Retrieved January 27, 2016, from http://www.nrdcactionfund.org/updates/dirtydenier-day-20-senator-james-inhofe.html/

NOAA Carbon Cycle Science. (n.d.). Retrieved February 8, 2016, from http://www.esrl.noaa.gov/research/themes/carbon/

NOAA Greenhouse Effect. (n.d.). *National Oceanic and Atmospheric Administration.* Retrieved January 22, 2016, from http://climate.nasa.gov/causes/

NOAA Water Vapor. (n.d.). *National Oceanic and Atmospheric Administration.* Retrieved October 2015, from https://www.ncdc.noaa.gov/monitoring-references/faq/greenhouse-gases.php

PEW - Energy Race. (n.d.). *Who's Winning the Clean Energy Race?* Retrieved March 10, 2016, from PEW Charitable Trusts - Research and Analysis: http://www.pewtrusts.org/en/research-and-analysis/reports/2014/04/03/whos-winning-the-clean-energy-race-2013

Princeton University, Climate and Energy Research Challenge. (n.d.). Retrieved October 2015, from http://www.princeton.edu/pei/grandchallenges/research/energy/

Reich, R. (2015). *Saving Capitalism.* Knopf.

Romm, J. (2015). *Climate Change: What Everyone Needs to Know .* Oxford University Press.

Scripps Institution CO2. (n.d.). *The Keeling Curve.* Retrieved February 16, 2016, from https://scripps.ucsd.edu/programs/keelingcurve/2013/07/03/how-much-co2-can-the-oceans-take-up/

Small Hydro International Gateway. (n.d.). *Small Scale Hydropwer.* Retrieved March 10, 2016, from http://www.small-hydro.com/about/small-scale-hydrpower.aspx

SOLAR FAQs. (2006, April 20). *Sandia National Laboratories.* Retrieved October 2015, from http://www.sandia.gov/~jytsao/Solar%20FAQs.pdf

Stephen Cass, M. T. (2009, August 18). *Solar Power Will Make a Difference.* Retrieved   October   2015,   from http://www.technologyreview.com/article/414792/solar-power-will-make-a-difference-eventually/

Suzanne Goldenberg, M. J. (2015, July 9). *Exxon Knew.* Retrieved November 2, 2015,   from   Mother   Jones: http://www.motherjones.com/environment/2015/07/exxon-climate-change-email

Suzanne Goldenberg, T. G. (2015, July 8). *Exxon knew.* Retrieved November 2, 2015,   from   The   Guardian: http://www.theguardian.com/environment/2015/jul/08/exxon-climate-change-1981-climate-denier-funding

U.S. Department of Energy - Hydrate. (n.d.). *New Methane Hydrate Research.* Retrieved March 10, 2016, from http://energy.gov/articles/new-methane-hydrate-research-investing-our-energy-future

U.S. Department of Energy Fossil Energy Source. (n.d.). *Fossil Energy Sources.* Retrieved March 10, 2016, from http://energy.gov/science-innovation/energy-sources/fossil

U.S. Department of Energy: ALGAE-TO-FUEL. (n.d.). *ALGAE-TO-FUEL.* Retrieved March 26, 2016, from Office of Energy Efficiency & Renewable Energy: ENERGY 101: http://energy.gov/eere/videos/energy-101-algae-fuel

U.S. Energy Information Administration. (n.d.). Retrieved October 2015, from http://www.eia.gov/environment/emissions/co2_vol_mass.cfm

U.S. Environmental Protection Agency - Oceans. (n.d.). *Climate Change Indicators, Oceans* . Retrieved March 10, 2016, from ·https://www3.epa.gov/climatechange/science/indicators/oceans/

U.S. Environmental Protection Agency. (n.d.). *Overview of Greenhouse Gases.* Retrieved   October   2015,   from http://www3.epa.gov/climatechange/ghgemissions/gases/ch4.html

UCS - Internal Fossil Fuel Industry Memos. (2015, July). *The Climate Deception Dossiers, Internal Fossil Fuel Industry Memos.* Retrieved November 2, 2015,   from http://www.ucsusa.org/sites/default/files/attach/2015/07/The-Climate-Deception-Dossiers.pdf

UCS Renewable Benefits. (n.d.). *Benefits of Renewable Energy Use.* Retrieved March 10, 2016, from http://www.ucsusa.org/clean_energy/our-energy-choices/renewable-energy/public-benefits-of-renewable.html#.VuMnh4-cF7c

US Dep Energy CO2 EOR. (n.d.). *Enhanced Oil Recovery.* Retrieved February 16, 2016, from http://energy.gov/fe/science-innovation/oil-gas-research/enhanced-oil-recovery

US Environmental Protection Agency. (n.d.). Retrieved October 2015, from http://www3.epa.gov/climatechange/ghgemissions/global.html

UWOSH Carbon Bubble. (n.d.). *Financial Analyses of Stranded Assets & the Carbon Bubble.* Retrieved Feb 16, 2106, from http://www.uwosh.edu/es/climate-change/divestment/carbon-bubble

WWF - Renewable Energy. (n.d.). *Will developing countries lead the renewable energy race?* Retrieved March 10, 2106, from http://climate-energy.blogs.panda.org/2015/10/08/will-developing-countries-lead-the-renewable-energy-race/

Yale Carbon Capture Tech. (n.d.). *Environment 360.* Retrieved February 16, 2016,   from http://e360.yale.edu/feature/can_carbon_capture_technology_be_part_of_the_climate_solution/2800/

Made in the USA
San Bernardino, CA
16 May 2016